How to Catch Lies

with Statistical Distributions

Table of Contents

Preface

Along with the certainty of death and taxes, there is one more certainty that every citizen faces – the fact that he or she is being lied to. Lies are thrown in the face of the ordinary people every day by the politicians, entertainers, businessmen, religious leaders, lobbyists and many other individuals and organizations. Anyone with vested interest in having people adopt a specific viewpoint has a strong incentive to lie. And many of these lies are disguised as statistics or scientific studies.

There are some excellent books that point out the different ways in which statistics can be used to misrepresent information and facts. These books reveal the different nuances and flavors by which statistics can be used to misrepresent and distort information. Some of the standard tricks pointed out in those books include selecting a convenient measure of the average (i.e. choose between mean, median and mode), select the axes of a chart to make a misleading impression, or select a type of chart to make a misleading impression. Thanks to those books, the ability of the vested parties to fool the public is much reduced

However, there are other ways to lie and misrepresent information to shape people's opinions. And some of these lies are told in such a subtle way that it is difficult for people to see the trick that underlies them.

And this is where statistical distributions enter the

picture. Many of the lies and misconceptions can be laid bare in a few minutes if you apply the concept of statistical distributions to test their validity. A statistical distribution may sound like it requires a Ph.D. to understand, but in reality it is a simple concept that anyone can comprehend and apply without the use of Greek characters.

Most of us understand the concept of distributions intuitively. When we are told that someone is tall, we understand implicitly that there is a distribution of heights and a person is positioned towards one end of that distribution. We can understand where a person lies in a distribution without knowing the specific details of the height. With many of the adjectives we use, e.g. tall, thin, fat, hot, rich, poor etc., we intuitively associate a distribution.

In some cases, expressing distributions in conversational language is somewhat cumbersome. We would express a statement like Americans are richer than Mexicans, which is a true statement if you consider that for the average of the income distribution of households in the two countries. However, it can be easily misconstrued to mean that any American is richer than any Mexican – which is obviously not correct. Yet, much of conversations and comments on various websites, web-blogs and radio talk shows are nothing but variations of very similar arguments. The difference in what we want to convey about distributions, and what we end up saying in natural language creates a tremendous opportunity to distort reality.

It is the intention of this book to expose some of those misconceptions and lies, and to explain enough about statistical distributions so that the reader will be able to discern similar lies when he or she encounters them in the future.

Not all of the lies are told with an evil intention. Sometimes, it is better to believe a lie than to know the truth, e.g. it is a bitter pill for any scientist to point out a flaw in the arguments that evolutionists make against creationists. Some of the lies are undertaken with a very good intent by some very scrupulous and honest leaders. Some of the lies are nothing but clever marketing that many organizations indulge in. Nevertheless, a misconception which is not grounded in reality needs to be pointed out and corrected. Any good and worthwhile cause can be supported without resorting to incorrect reasoning.

Although there are a few simple techniques that are used to create the misconceptions, these can be used in many creative combinations to generate new lies. The book follows the approach of identifying a few common lies/misconceptions, points out the flaw in their reason and discusses how statistical distributions can be used to readily expose them. Each chapter of the book covers one such misconception or lie in detail.

In order to prevent similar types of lies and misconceptions, you also need to understand the general means by which these lies are construed. The first and last chapters of this book provide those general concepts. The first chapter provides an overview of statistical distributions. This is followed by chapters that examine one specific instance of a lie or misconception in the society, and exposes the flawed reasoning underlying it. The last chapter summarizes the generic means for catching lies using the concept of statistical distributions.

৵৹৶

1. A Brief Introduction to Distributions

The concept of a statistical distribution is fairly simple and you may recall it from some of your mathematics courses at school. If you try to measure some property among a set of people or objects, you are not likely to get a single value but a set of values. How these different values appear is a statistical distribution.

As an example, suppose you stand at the intersection of Meridian and Michigan streets in Indianapolis (or any other place in your local city where a lot of your acquaintances are likely to pass through), and measure the height of the first ten of your acquaintances that you see passing by. You want to restrict this experiment to your acquaintances since others may give you a strange look if you actually did try to measure their height, and in some localities the local police department may decide to intervene and stop this experiment. If you are shy, or do not have many acquaintances, just estimate a person's height within the accuracy of an inch by eyeballing them. You will get ten different measurements for the heights of different people you come across. Instead of getting a single value, you get values that are distributed across a range. What you have just collected for yourself is a statistical distribution.

The set of ten values you get in this experiment may look like a collection of random numbers. However, since most of the things we measure in real-life form a statistical distribution, statisticians and mathematicians have come up with some metrics to characterize the data that is collected in this manner.

Some of the simple metrics that everyone is familiar with are the minimum and maximum of a set of data values, the range of the data, the mean of the data, median, mode, quartiles and percentiles. The range is the difference between the maximum value and the minimum value and the mean of the data which is the mathematically averaged value of all data that is collected.

The median is the value which is exactly half-way across all the collected measurements if you order them in an increasing order. The first quartile is the smallest value which is greater than a quarter of measurements, and the third quartile is the smallest value which is higher than three quarters of the measurements. The second quartile is nothing but the median. The first percentile is the value which has 1% of measurements below it, the twenty-first percentile is the value which has 21% of the measurements below it, and so on. Of course, you will get meaningful percentile values only if you were collecting a large number of measurements. A percentile across a set of ten measurements does not quite make much sense unless you are talking of the 10th or 20th percentile. But if you were measuring the heights of a thousand people who crossed the intersection of Meridian and Michigan, the second or third percentile would also be meaningful to talk about.

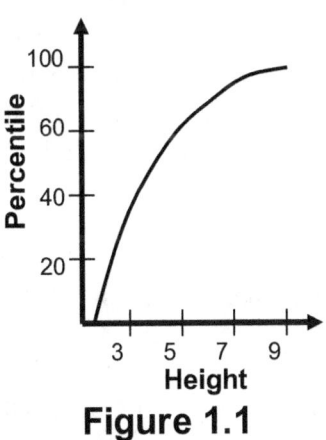

Figure 1.1

You could plot the set of all values you get in any measurements against the percentiles as a graph. You can draw it so that the heights are along the horizontal axis and the percentiles are along the vertical axis. Now you have a graph of the nature shown in Figure 1.1. You will have most of the people measuring between one foot (for babies) and eight feet (really tall friends), with the hundred percentile being somewhere near the eight foot mark and the median being somewhere around six feet.

If you repeated the experiment at the intersection of Meridian and Michigan for a number of days, you are likely to get a different set of measurements on each day. One day, a set of your neighbor's middle-schooler and his friends may walk by, while another day you will find the tall basketball player next door come by with his team of really tall friends. Such differences not withstanding, on most of the days you will find an approximate similar mixture of people, and you will find that the distributions you get on any two typical days have a lot of similarity. This holds true for many types of measurements that we experience in real life. If you kept on adding the observations made on each day, you will probably find that the distribution of all measurements made up till the 100th day and up till the 101st day are not very different from each other. Basically, as you keep on making more and more measurements, the plot of distribution you will make will change less and less as you increase the number of days you make the experiment.

The theoretical distribution you will get if you kept on measuring the height of all your acquaintances indefinitely is the cumulative probability distribution function (CDF) of the height of your friends[1]. It is very

---

[1] To be mathematically accurate, we needed to take care of some nuances such as eliminating counting the same friend twice as we made the measurements. To simplify the discussion, all those

similar to the graph in Figure 1.1 except that the vertical axis is labeled with a number between 0 and 1 instead of a percentile. The vertical axis shows the statistical concept of probability. A probability is a measure of the chance of something happening, with a probability of 0 meaning something has no chance of happening, and a probability of 1 meaning that something is almost certain to happen.

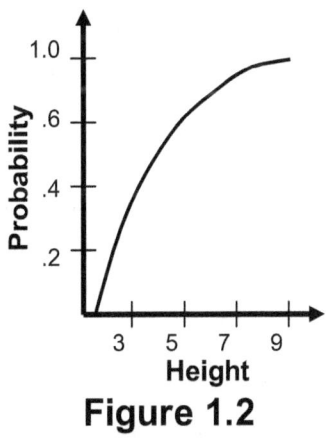

Figure 1.2

Figure 1.2 shows a CDF you might get from the experiment at the corner of Meridian and Michigan. If the graph shows the probability value of 0.6 for a height of 6 feet, it means that there is a 60% chance that any randomly selected friend is less than 6 feet tall. It also

---

nuances are ignored in this introduction.

means that for all intents and purposes, 60% of your friends are shorter than 6 feet. The caveat of "all intents and purposes" is made because some pathological cases can be constructed when this interpretation of probability will be technically incorrect.

Another way to interpret the probability shown on the vertical axis in Figure 1.2 is to consider it as the percentile value that will be obtained when measurements are taken over a very large set of people, in theory an infinite number of people. In practice, if you are measuring the heights of a few hundred people, you are highly likely to come with a distribution that is reasonably close to the theoretical one.

An equivalent representation of the same distribution can be drawn where the probability on the vertical axis is not a measure of the chances of something being lower than a point on the curve, but of the probability that the value is equal to the point on the curve. That curve is called the probability distribution function (PDF), as opposed to the cumulative probability distribution function (CDF). While the CDF grows from a value of 0 to a value of 1, the PDF always is between 0 and 1 but grows and shrinks between these two limits. The PDF has the property that the area bounded by the horizontal axis and the PDF curve is always 1. The probability distribution graph gives some information that is easier to see visually than the cumulative probability graph.

A sample probability distribution function is shown in Figure 1.3. From this graph you can easily figure out the height which is most common among your friends, information that is not that obvious from the cumulative distribution function. Of course, you can also calculate it using a variety of mathematical constructs. In that respect, the information contained in the CDF curve is the same as the information contained in the PDF curve.

The probability distribution function is what we will be using in most of the discussions in the rest of the book. It visually shows how the measurements made of any quantity are distributed.

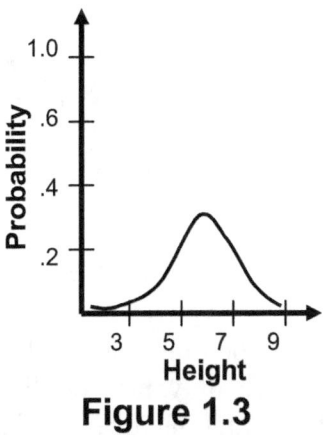

Figure 1.3

Most of the things that one can measure quantitatively in any social, economic or political context can be represented as probability distributions. If you consider the average income of American households, that will be a distribution with some PDF. If you consider the weight of American kids between the ages of 8 and 12, that will be a distribution with some PDF. If you decide to rank the patriotism of all Americans on a scale from 1 to 10, you will get a PDF. If you decide to rank the quality of telephone calls on a scale from 1 to 5, you will get a PDF. These PDFs and distributions are all around us.

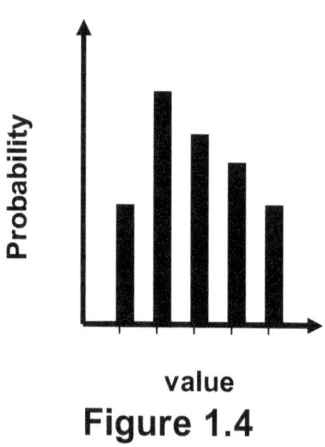

value

Figure 1.4

A PDF can be drawn when the horizontal axis has continuous range of values or discrete values. As an example, if you rank the quality of calls you make on a continuous scale from 1 to 5, you will get a curve that looks like one of Figure 1.4. However, you can also be in a situation when you are only allowed to rank them on an integer scale, i.e. a value of 2.5 will not be allowed. You can only choose 2 or 3 as the closest option among a value from 1 to 5. In those cases, the PDF would look like a bar graph instead of a curve, as shown in Figure 1.4. The heights of all the bars will need to add up to 1. Such a PDF is known as a discrete PDF as opposed to a continuous PDF.

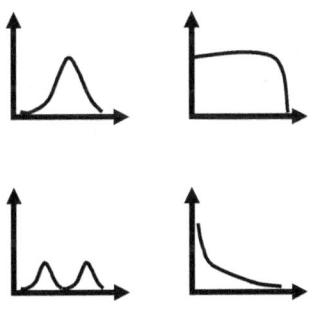

Figure 1.5

In theory, a PDF can have any shape, and Figure 1.5 lists some of the shapes that are possible for the PDF of different types of continuous measurements. In general, one can imagine any type of shapes of the PDF. The measurements you make under different conditions are likely to have a very different PDF. However, most of the common distributions that we will encounter in our real life tend to follow a PDF that is known as the Normal distribution.

The Normal distribution has a PDF like the one shown in Figure 1.6. It has a peak in the middle, and it decreases in a symmetric manner on both sides of that peak. The value corresponding to the peak is the average value one would expect of the quantity being measured, and how fast the curve decays is a measure of how dispersed the values are around that average. If the values are clustered closely around the average, the curve will fall down very rapidly. If the values are more dispersed, the curve falls down more gradually. In the Normal distribution, as you move further away from the mean, the probability decreases rapidly, and it is close to zero for values that deviate far from the average.

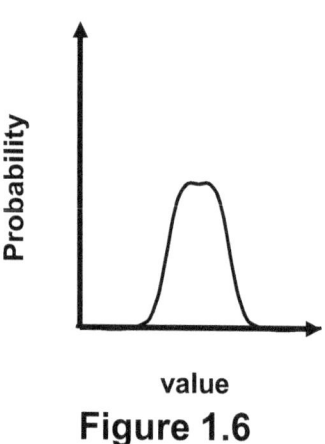

value

Figure 1.6

There are other types of statistical distributions, but for the items discussed in this book, it is okay to assume that the values we are measuring will correspond to a

Normal distribution. Most of the things we will encounter in life follow some type of distribution and approximating them to the Normal distribution is not too bad of a distortion. And for the type of qualitative discussion around catching lies that we are discussing in this book, making that assumption would not have us reach any incorrect conclusions.

Although most things in life are distributions, it is hard to describe and talk about distributions for the different values. It is much easier and simpler to talk about a single value, such as the average of a distribution. This gives a lot of opportunity for the demagogues to mislead the public by being selective about how they talk about any of the measurements. And it is not just the demagogues. Even ordinary citizens make improper generalizations of this nature.

Beginning from the next chapter in this book, let us look at some of the common distortions and misconceptions that are laid bare if we just apply the concept of statistical distributions.

2. Bigots by Chance

If you have watched the movie "Terms of Endearment," you might recall the scene where Sam Adams chides a checkout girl for being rude to Emma Horton. When the checkout girl protests that she was not being rude, Sam retorts, "Then, you must be from New York." The movie was excellent, but this dialogue did portray an opinion about the residents of New York that was somewhat bigoted.

New Yorkers would have been justified in taking offense at that dialogue. In July 2006, Reader's Digest published a survey of politeness and good manners in about 35 large cities across the globe, and New York came out at the top in civility and politeness. The survey had a disclaimer that it was not conducted scientifically, but let us use that as an informal measure of politeness in cities anyway. Let us go down the list to the European or North American city that was ranked lowest in the survey. That city would be Bucharest, Romania. We are skipping South American, Asian and African cities because the checkout girl in the movie could not have been expected to come from those regions due to her ethnicity. Would Sam Adams have been justified in saying to the check out girl that she must have been from Bucharest?

The answer is a definite no. No matter which city name was used in the dialogue, it was a bigoted statement. Even if the survey of Reader's Digest was done in a scientific manner, and we assume that there is enough justification to believe that a higher percentage of people in Bucharest (or New York City) are ruder than in other cities, it is incorrect to say that the checkout girl must have been from that city just because she made a rude comment. That statement was incorrect, bigoted and stereotypical, even though it made a very good dialogue for that particular movie scene.

There is plenty of similar bigotry and stereotyping that goes on in our daily lives. My son likes to play soccer, and a large part of my weekend is spent driving him over to the matches organized by AYSO – the American Youth Soccer Organization. In the series of one of those matches, we ended up playing against a neighboring town which has a slightly different mixture of ethnicities than ours. When the teams came out on the field to practice, the contrast was striking. While our town's team had mostly white players, the other town's team had four players who were obviously Hispanic. One of the other parents in my teams leaned over and whispered, "We are dead meat. Those Hispanic kids will run through our defenses like they are made of butter." He was taking it for granted that the Hispanic kids will be better at soccer than the white kids in our team.

I do not know if the brown complexioned children on the other team were really Hispanics, but let us just assume that they were. It is true that several teams from Hispanic countries in Latin America have a much better standing in international soccer than any American team. Nevertheless, it is blatantly incorrect to assume that a team with Hispanic kids will trounce a team with white kids from that.

The mistake made in both of these examples is confusing the properties of an individual with the properties of a group. A property of a group will be a distribution, while the property of an individual will be a sample or a single point in that distribution. And just because two individuals are drawn from two different distributions do not mean that they can be compared like their distribution.

To make it more concrete, let us consider the distribution that will be seen if we were able to actually measure the rudeness of people in a city. We can plot the PDF (probability density function) of rudeness of the folks in a particular city in a manner that is shown in Figure 2.1. In this figure, the vertical axis marks the probability. We assume that rudeness is plotted on the horizontal axis, and that the measure of rudeness increases as we move along the horizontal axis. Let us also assume that the distribution of rudeness in any city follows a Normal distribution. Although it is highly unlikely that any researcher will get a grant to exactly plot the nature of the distribution, the characteristics we are assuming will not be unreasonable to expect as a characteristic of any large population.

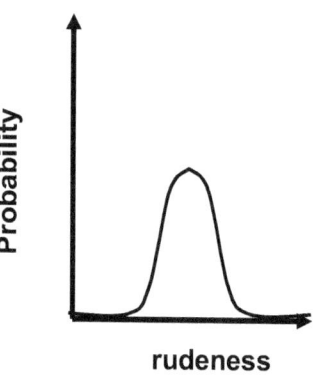

Figure 2.1

The picture is not complete yet. While we are assuming that any two cities have the same shaped PDF when you look at the rudeness of their residents, it does not follow that the PDF is exactly the same. Let us take a look at the cities involved in the "Terms of Endearment," namely a small town in Iowa and the big metropolitan area of New York City. They would have the PDF of rudeness as shown in Figure 2.2. The Iowa curve is on the left on the New York City curve showing that on average a small town resident is more polite than the resident of the big city. A similar result would be expected if you compared any two cities in the Reader's Digest by plotting the PDF of the rudeness of their residents.

Now, let us take any two residents in the two different cities and try to compare the relative rudeness of the two. Is it true that the person from Iowa will be more polite than the person from New York City?

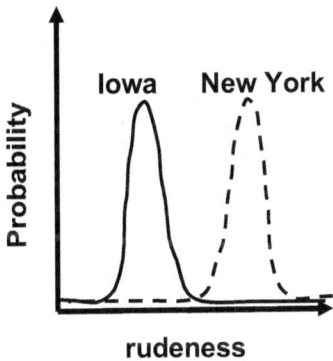

Figure 2.2

If we are looking for a chances or likelihood, it is correct that a randomly selected person from Iowa will likely be less rude than a randomly selected person from New York City (assuming of course that Figure 2.2 depicts reality). However, it is not correct to assume that if you take any New Yorker and any Iowa resident the Iowan will be less rude.

A simple counter-example is provided in Figure 2.3, which shows a person A from New York and a person B from Iowa. As shown in the figure, person B is ruder, even though he (or she) comes from a city where people are less rude in general. You can not compare the specifics of two individuals by inferring it from the general distribution you expect their groups to have.

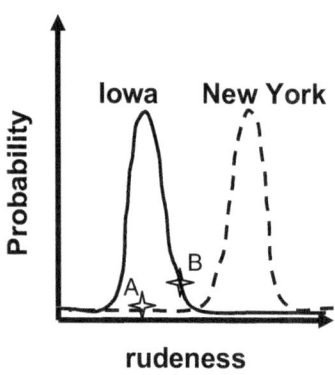

Figure 2.3

So, it is unfair to brand a checkout girl in Iowa to be coming from New York just because she is rude. However, as a retort to put down people, one must admit that the dialogue was a snappy one, even if it was incorrect.

The same holds true for the soccer skills of the white and Hispanic population in the United States. It is not obvious why just being Hispanic makes an American kid be better at soccer. But let's assume that the average Hispanic kid is a better soccer player than the average white kid. We will again assume that we can quantify the skill at soccer and measure it across the population of white and Hispanic kids in the US. We will get a diagram like than in Figure 2.4.

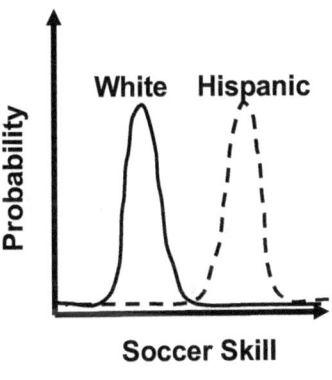

Figure 2.4

The PDF of soccer skills of the white kids peaks at a skill level before the PDF of the Hispanic kids. However, if you select individuals from the sample, you can very well end up with a situation where a specific white kid is better than the Hispanic kid at soccer. The parent on my son's soccer team transferred the attributes of a group to the individuals. In that specific case, it was lucky for our team that he was wrong.

There is some justification behind the statement made by my fellow soccer team parent. If one randomly picked a Hispanic kid, and randomly picked a white kid whose soccer skills were distributed according to the PDF shown in Figure 2.4, one would expect that the chances would be higher that the Hispanic kid would have better soccer skills than the white kid. How much the chances are would depend on how far apart the two PDF functions were.

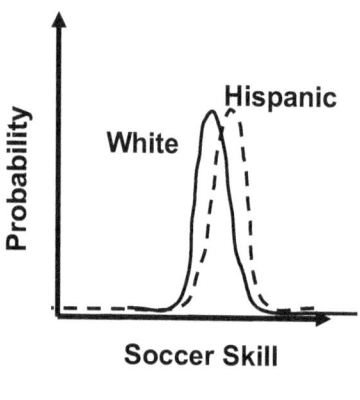

Figure 2.5

The main difference between the PDF of the Hispanic kids and the white kids is that one is shifted right. If the shift is just minor, then the probability that a randomly picked white kid is better than a randomly picked Hispanic kid is higher. That would be the situation depicted in Figure 2.5. The peak of the Hispanic distribution is at a skill level which is slightly higher than that of the white kids. And while this shift allows one to say in general that on the aggregate Hispanic kids are better at soccer than white kids, it says virtually nothing at all about the relative merit of one individual Hispanic kid against one individual white kid.

Of course, it is a very different situation where the PDFs are far apart. Assume that the PDF distribution of soccer skills were as shown in Figure 2.6. If the distribution were really so far apart, then the chance of a randomly picked white kid to be a better soccer than a randomly picked Hispanic kid will be negligible (not zero but negligible). In that case, it was a miracle that our team won against the team with Hispanic kids.

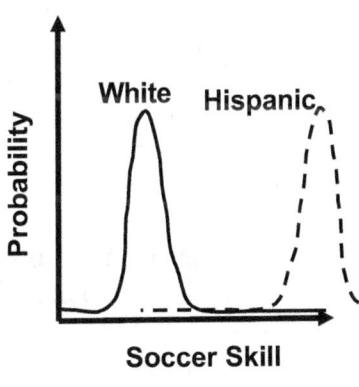

Figure 2.6

There is no scientific study of soccer skills by ethnicity, nor is one likely to be there anytime soon which will give us an exact quantification. If I have to guess which one is a more likely distribution of soccer skills, I would pick Figure 2.5 over Figure 2.6. In my unscientific study of soccer players, based on nothing more than the limited sample of teams my son has played against; it is a better fit to real life.

A significant amount of bigotry is introduced by our natural tendency to generalize statements about groups of people. Let us assume that Figures 2.4 through 2.6 are not about the relative soccer skills of Hispanic and white kids, but about the relative basketball skills of white males versus black males. Based on prevailing stereotypes, one could map the dotted PDF to the skills of black males and the solid PDF as the skills of the white males. But it is cumbersome to describe the skills as PDF, and it is so succinct to summarize the comparison in a phrase like – "white men can't jump." An entire movie has been made around the misconception that happens when one forgets that some statements are about comparing distributions and can not be used to compare individuals.

I had come across an amusing scenario in the sports club that I belong to. The club organizes league matches for several sports, and the tennis league matches are particularly popular. I came across a poor teenager who was being mercilessly taunted by his friends for having lost to a girl in a league match. The poor victim was protesting to no avail that the girl in question was a more experienced player and so the loss was perfectly fine. His friends were having none of that. If we compare the relative tennis skills of male and female members of the club, we will immediately understand why that was so. The situation is another instance of the situation demonstrated in Figure 2.3. While a distribution of tennis skills of boys versus girls is likely to show the PDF of boys to peak at a higher level of skill than the PDF of girls, one can easily have situations where a specific girl is better at tennis than a specific boy.

It is not just ordinary citizens who make this type of generalization and draw incorrect conclusions. There have been some famous incidents of such generalization with very visible though unfortunate results. One of the incidents which could have been avoided by improper generalization was that of Juan Williams. Juan was a very successful host on National Public Radio. In an interview in October 2010, he made a statement in a television interview that he got very nervous when he saw someone in a Muslim garb flying at an airport. This statement led to NPR firing him.

Let us examine whether Juan was right in making his statement. We can draw the PDF distribution of terrorist tendencies among the Muslim versus the non-Muslim population in the United States. Let us assume that the Muslim population tends to have more terrorist tendency, and justify that on the fact that several of the incidents of terrorisms that have occurred on United States soil have been due to individuals who were Muslim. In that case, we are likely to get a PDF distribution of terrorist tendencies which look like that shown in Figure 2.7. The dotted line corresponds to the PDF of terrorist tendency in the Muslim population, and the solid line shows that of the other non-Muslim population.

Shifting the curve for Muslim population to the right may itself be viewed as a racist or bigoted action. There is no survey or study that shows that Muslims in the United States have more terrorist tendency than the non-Muslims. However, let us just assume that to be true, and see where it leads. If you are a Muslim reading this book, please hold your anger in check until you read a few more pages in the book.

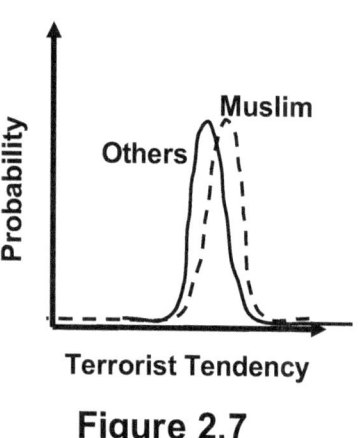

Figure 2.7

If we see a person who is clearly Muslim at the airport, we are talking of an individual from the dotted distribution. And if we pick a non Muslim person at the airport, we are again talking of a specific individual from the solid distribution. And we can easily find instances where the terrorist tendency of the non-Muslim would be higher than that of the Muslim. Just think of the terrorists that bombed the federal buildings in Oklahoma City.

The basic problem in Juan's reaction to seeing a Muslim at the airport was applying the properties of a distribution to that of an individual. If the shift in the PDFs of the two distributions is minor, it is unfair to apply that generalization and conclude that a specific Muslim individual has a higher terrorism tendency. If you try to guess the shift in distribution between Muslims in the United States versus the non-Muslim population, it is likely that the shift will be relatively minor. On the other hand, if it was a comparison of the United States population to the Muslim population in a fundamentalist society, e.g. the areas on Afghanistan Pakistan border where the Taliban still have a strong presence, the shift in the PDF positions could be expected to be quite large, and there will be more of a justification for Juan's position.

So, Juan was wrong in trying to apply the properties of a distribution to an individual. I wonder if he would still have his job if he explained his unease in terms of probabilities and PDF instead of an absolute statement about seeing a specific person in a type of garb.

It is not just ordinary people who use such type of generalization. Some people with very deep technical expertise and mathematical knowledge also fall into the same trap. In my regular job, I manage a team of professionals who all graduate from college with a full slate of courses, including several courses in mathematics and statistics. The company I work for assigns employees to levels, each level associated with different responsibility and a different salary-range. An employee in a lower level would have lower set of responsibilities than an employee in a higher level, and generally a lower salary.

In one year of my career, I ended up hiring two employees. One (employee A) was from an college that was ranked among the top 20 colleges in the United States while the other (employee B) was from a college whose rank was in the 50s. However, the employee from the lower ranked college had a higher number of individual accomplishments. As a result, B was hired into a higher level than A.

People talk to each other at office, and when employee A learnt that employee B had a higher level, he was livid with rage. He marched into my office and demanded a change in his level. His main argument was that he was from a higher ranked University than employee B, and thus entitled to the same or higher level. I listened patiently and then drew the diagram shown in Figure 2.8 on the board for him.

I then pointed out that he was at position A based on our interview and the other employee was at position B. While the PDF curve of skills of graduates from his University was shifted to the right compared to those from the lower ranked University, it did not automatically make him better than everyone in the lower rank University.

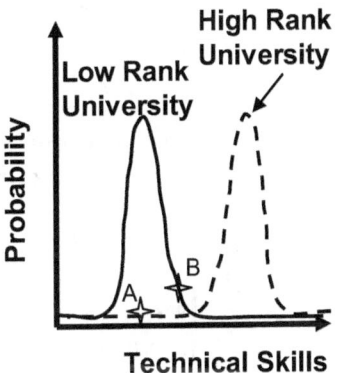

Figure 2.8

The gentleman thought over this for a few minutes in silence and then left my office. He has been a model employee ever since.

Another type of common confusion that often results in bigoted view is when people try to replace the distribution with a sample, or when a single point is taken as a representative for the entire distribution. Let us take the case of Bernie Madoff, who is the infamous investment advisor who deceived several prominent people using a Ponzi scheme. The fall-out from his exposure was wide-reaching. In the light of this scam, various statements were made in the media including statements like "All investment brokers are crooks" and "All investment brokers are out to swindle you out of your money." These statements usually were made by the irate members of the public, and you can still find similar comments on various blogs and websites.

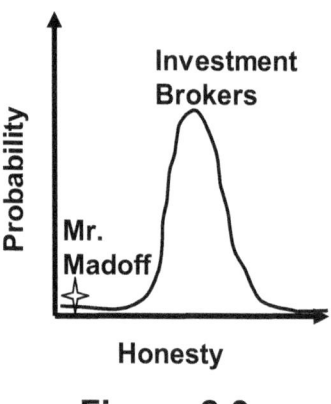

Figure 2.9

If you look at the honesty of investment brokers, there will be a distribution in how honest they are. Let us say that the distribution looks like Figure 2.9, with Bernie Madoff being located at the extreme left of the curve.

However, generalizing from that one point to the case where all investment brokers are crooks is trying to distort the shape of the PDF so that it looks like that of Figure 2.10. In that figure, the entire PDF is drawn so that most of the brokers are concentrated in a narrow zone around Mr. Madoff. However, that is just not correct to condense a large distribution to one point.

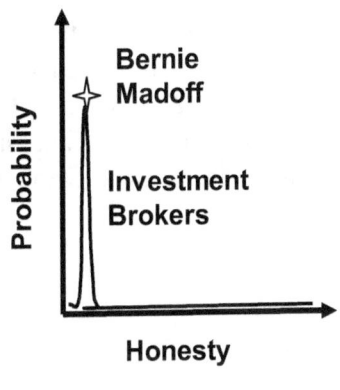

Figure 2.10

Unfortunately, the phenomenon of using one or a few points to collapse an entire distribution is not too uncommon. It is more convenient to speak as a gross generality rather than in terms of a distribution. As an example, let us consider the following statement from United States Attorney General Erich Holder. In a speech to Justice Department employees marking Black History Month, Holder said the workplace is largely integrated but Americans still self-segregate on the weekends and in their private lives.

In a country of 250 million and more, there will be a spread in how much integration or segregation happens in private lives of people. It would have a spread like that of Figure 2.11.

There are many Americans who are completely

integrated in their private life, and it is quite likely that there are some Americans who prefer to mix with only people of their own ethnicity. Mr. Holder probably meant that there are still many Americans who prefer not to integrate in their private life. That statement would be consistent with what we expect and know of American life in general.

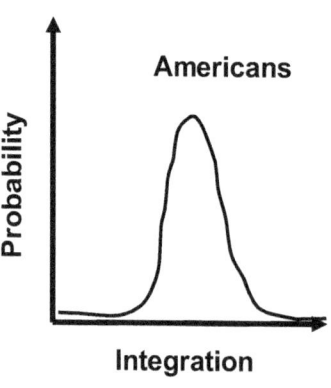

Figure 2.11

However, what Mr. Holder said could also be interpreted as stating that the distribution of Americans in the context of integration looks similar to Figure 2.12. In that figure, the PDF of integration is remarkably shifted to the left, showing that Americans do not integrate in personal life at all, or that the integration has not happened.

There have been no large-scale studies of

integration on which the exact nature of these curves could be quantified. Few people will disagree with Mr. Holder's assessment if he meant Figure 2.11. An interpretation of Mr. Holder's statement as being Figure 2.12 may not be an appropriate representation of reality.

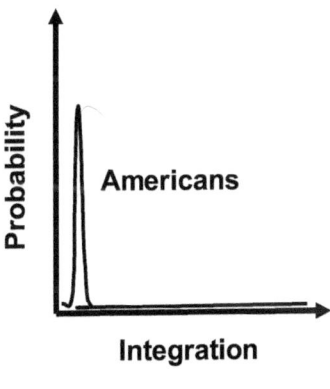

Figure 2.12

3. The Mythical Asian Genius

Have you seen the movie "Monsters versus Aliens"? It is a delightful science fiction animated movie with many hilarious scenes and funny dialogs. In the early parts of the movie, when an alien robot attacks the United States, the president asks his staff to get scientists working on finding a solution. The advisor to the president barks out an order, "We need our best minds on this. Get India on the phone." The movie was poking fun at the high percentage of Indians in various science and technical fields, along with the business focus on outsourcing different types of jobs outside the United States.

In popular media, you find portrayals of Asians (including Indians, Chinese, Japanese and other East Asian/South Asian/South East Asians) as geniuses, academically gifted, and usually nerds. In the movie "Mean Girls," two out of five people on the Mathlete team are Asians. In the television series "Phineas and Ferb," the Asian Indian character of Baljeet is a studious geek. In the television series "Grey's Anatomy," the Asian character of Christina Yang is always striving to be the best, at top of the class in Stanford, and extremely competitive. In another television series *"Hey Arnold,"* the half-Japanese character of Phoebe Hyerdahl is a studious and competitive geek. In the immensely popular cartoon strip of *"Dilbert,"* the Asian Indian character of Asok is capable of performing miracles because he comes from Indian Institute of Technology (IIT). You can find many similar stereotypes in popular media depicting Asians as studious and geniuses.

And there seems to be enough evidence in real world to justify that media portrayal. Look at the winners of the United States Scripps National Spelling Bee over the last several years. The winners for the four years of 2008 to 2011 have been Asian Indians. If you look at the roster of the top ten or top forty contestants in any year, the percentage will be largely Asian. If you drop by any of the Mathematics school competition at the state level in almost any state, you will find yourself surrounded by several enthusiastic Asian faces. And Asians make up a disproportionate number of engineering and scientific graduate programs of the major Universities. One of the professors from a leading West Coast University had quipped in half-joke that their graduate program admits one third of their applicants from India, one third from China, and the remaining third from other Pacific Rim countries including the United States.

Is there something special about the Asian people that make them better in science, mathematics and technology than white people? Are they somehow superior to other races like whites, blacks and Hispanics in these skills?

The answer of course is No. The race of an individual has nothing to do with their skills in mathematics, arts or sciences. Several other factors do influence these abilities, but race/ethnicity is not one of them. But then, how do you explain the preponderance of Asians in all of the highly specialized fields?

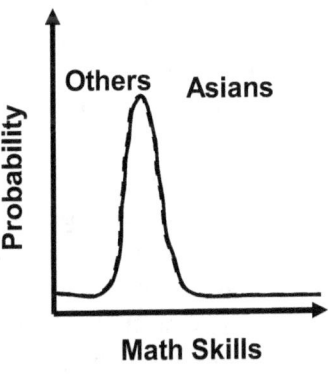

Figure 3.1

Let us consider mathematical skills as a proxy for difference kind of technical expertise in various fields. Once again, the answer is fairly obvious if you look at mathematical skills through the means of statistical distribution. Let us consider the distribution of any skills among a population. It would tend to follow the Normal distribution that we have been using throughout this book. If you take any population, either Asian or non-Asian, you would expect to have the same distribution of mathematical skills. It would be a Normal distribution like that shown in Figure 3.1.

Mathematical ability is dependent on many factors, some based on innate factors such as genetics, and other based on external factors such as education, access to libraries, social stability etc. which allows one to improve and apply the innate mathematical skills. Let us assume that the external factors are the same for all ethnicities in any given environment. Then it is natural to expect all ethnicities to have a similar distribution in their innate mathematical abilities.

Note that the same would not hold true if the external factors were changed. If a country has widespread poverty, poor schools and library system, and a poorer social infrastructure, its people would have a lower mathematical skills than the people in a country with a higher income level, better education system, and a better social infrastructure. Let us consider the country to which the American President's staff turns to get the best technical minds in "Monsters versus Aliens," namely India. If we were to take the PDF of actual mathematical skills across the entire population of India, and compare that to the technical skills of the entire population of the United States, one would expect to get a distribution like that in Figure 3.2.

The reason for this relative positioning of the two distributions would be that the GDP per capita of India is only $1,200 per year, about 2.5% of the United States GDP per capita of about $47,000 per year[2]. The literacy in India is 64% while the literacy in United States is 99%. Without exposure to good education, a poor literacy rate, and with significantly less money to spend on education pursuits, the actual mathematical skill distribution in India is likely to be much lower than that in the United States. The innate mathematical skills distribution of the Indian population is the same as that of the innate mathematical skills of the American population, but environmental factors result in shifting the curve to the left for Indian population in general.

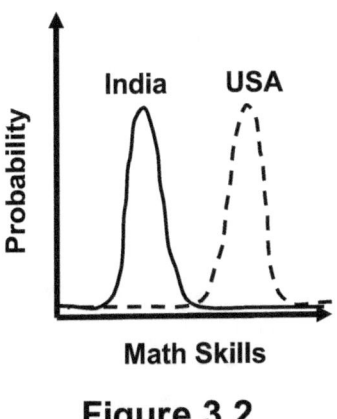

Figure 3.2

---

[2] The statistics have been taken from Wikipedia.

A similar situation is likely to be true for most of the Asian nations. The broader Asian population has a distribution which is shifted left in mathematical skills than that of the American population.

What then is the reason underlying the dominance of Asians in the technical and mathematical fields in the United States. Based on the observations made in the beginning of the chapter, it is fairly obvious that the Asian population in the United States has a much stronger skills in mathematics and other technical fields and other ethnicities. Do Asians have some innate advantage in technical skills, so that when they are placed in an environment comparable to that of the other United States residents, they excel in Mathematics and Sciences? If that is correct, the PDF of innate mathematical skills of Asians compared to non-Asians would be as shown in Figure 3.3. Is it indeed so?

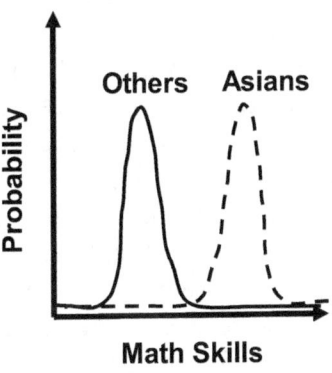

Figure 3.3

That would have been true if we do not take another factor into account, namely the immigration act of 1965. Until the Hart-Cellar act of 1965 was passed, immigration to the United States was limited by very stringent quotas on people of different national origins. For all practical intents and purposes, it eliminated any significant legal immigration by anyone but white Europeans. Once those admittedly biased restrictions were removed, more Asians started to immigrate to the United States. The Internet boom at the turn of the century also spurred a large boom in Asian immigration. If you look at the Asian population in the United States, you will find that a large percentage of them are first or second generation immigrants. Estimates of first or second generation immigrants as a percentage of the total population of the immigrants are invariably over 60%.

Who exactly emigrated from the Asian countries to the United States after the immigration policies were released? It definitely was not the poor villagers of India and China who did not have access to a good education. It was the type of people who are caricatured as Asok in the "Dilbert" cartoon strip. People who went to highly selective schools like the IIT in India, or the premier Universities in China, South Korea and Taiwan who were able to immigrate to the United States. These are people who had to have a significant amount of innate mathematical and technical skills, access to good education, and a stronger than average drive to be able to emigrate, adjust to the new country and find a job in a foreign land – which is only available if the employer can prove that their function can not be satisfied by any current United States citizen. That is a fairly exceptional set of people who will be at the very top end of their population. In other words, if you compare the Asians who immigrate to the United States against the Asians who don't, you will find the two distributions looking like Figure 3.4 [3]

---

[3] The emigrating population is small fraction of non-emigrating population. That is why the curve is relatively low in height.

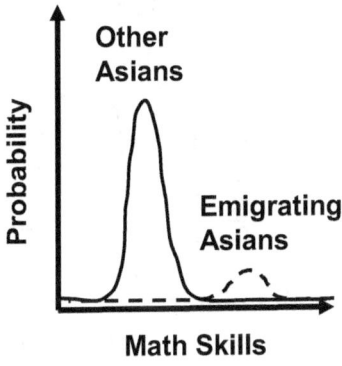

Figure 3.4

So, the sample of Asians in the United States is the very high end of their technical skills. You are taking a selected few from the Asian countries, and comparing them against the average population of the other ethnicities in the United States. If you now assume that the distribution of mathematical skills in the United States and overlay them on top of Figure 3.4, you get the distributions as shown in Figure 3.5. This relative position of the distribution will remain unchanged regardless of whether you use Figure 3.1 or Figure 3.2 as the distribution of mathematical skills between the United States and any Asian country. And in this comparison, it makes perfect sense that the average United States population (excluding the Asians) as compared to Asian immigrant population looks like Figure 3.3.

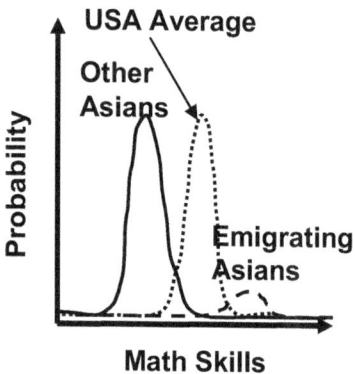

Figure 3.4

So, that settles the question. The American Asian population does better at the technical field because of the selective process they have gone through to come to the United States. The Asian races do not have any innate genetic traits that make them better at the technical fields than other races.

4. Putting Others Down

In the previous chapter, you saw an example where different portions of two distributions were compared against each other, creating a misleading impression about the superiority of one group over another. The fact that one could fudge around how different distributions are being compared is not lost upon many leaders who want to promote a strong cohesive mentality among their followers.

A common case of such fudging of information can be found in the duels that range on American radio waves among different branches of Christianity. For an instance of the same, let us select the virtual war that is waged by evangelical Christians against the Catholics. If you listen to the evangelical Christian radio shows, and there are many to choose from in every region of the United States, you will find them bashing Catholic thoughts every other day. And they had a heyday when the child molestation scandal of the Catholic Church became prominent.

In 2002, the Boston Globe covered a series of cases that involved sexual molestation of minors by Catholic priests. Ultimately, it became clear that there was a long history of such abuse and an attempt to cover it up by the Church establishment. This led to a significant outcry in different forums, with different radio shows blasting the Catholic Church as being the house of evil and with wide-spread condemnation of how they have let the system down. Invariably in those vitriolic speeches, the specific leader would state or insinuate that "this type of outrageous behavior will never be tolerated in church of blah" where blah stands for the Christian denomination that the radio talk show host belonged to. This statement was usually followed by some examples of the great acts of charity by the church of blah, followed by various phone calls of outraged church members praising the greatness of their group and expressing disgust at the abominable conduct of the Catholics.

An extremely amusing aspect in the treatment of this scandal that in all of the radio talk shows, the callers were always picking up from the best of their church members and comparing them to the worst of the Catholic Church. And in that comparison, the Catholic Church always came out looking the worse. Never mind the fact that there have been quite a good deeds by the Catholic people world-wide, and if one looked close enough, a sufficient number of evil actions could be discovered from any of the evangelical churches themselves.

We can come to the root of this unfair comparison by looking at the distribution of goodness of any group of people. Let us define a hypothetical measure of goodness which is very low for people who do all kind of evil deeds such as pedophilia, murder, stealing and other unspeakable actions, and very high for people who do all kinds of good deeds such as helping others, healing the sick, feeding the hungry, living according to good morals and the like. Let us also exclude from the goodness metric the requirement to belong to a specific religion or sect.

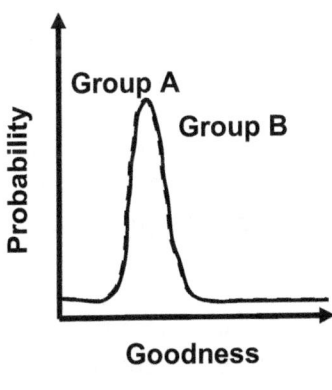

Figure 4.1

The exact definition of goodness can vary, but if the goodness metric does not involve the requirement to believe in the doctrines of a specific religion, sect or cult-leader, you would expect the PDF of the distribution of goodness to be more or less the same in any religious group. If you compare two religious groups A and B, the goodness PDF would end up looking like the diagram in Figure 4.1. That is a reasonable assumption based on the fact that apart from the choice of a religious group, the traits that make someone a good Catholic are the same as that which make one a good Episcopal or a good Presbyterian. And assuming that the definition of metric does not include the choice of the particular name of an omnipotent force (God versus Jehovah versus Allah) or a special leader (Jesus versus Moses versus Mohammad) these traits are not very different than the traits which make someone a good Christian or a good Jew or a good Muslim. There is also some statistical justification for this position. A study published in Policy Review of Hoover Institution at Stanford University in October 2003 found that the percentage of charitable donors made by members belonging to different religious groups were about the same, though larger than those made by atheistic groups.

The diagram of Figure 4.1 is not very palatable to the leader of most religious groups. It offers him no advantages in order to entice people to join his flock. The leader of any religious group, let us call it Group A, would much rather see the goodness PDF which looks like the diagram in Figure 4.2 when compared to that of any other religious Group B.

How can the leader of Group A motivate and influence his or her followers that the PDF of goodness are as shown in Figure 4.2? One obvious way is to include faith and devotion to the group's ideology in the goodness metric. In one fell swoop, that redefinition of goodness metric ensures that the goodness PDF of the members of Group A shifts to the very right while that of the other groups is shifted to the left.

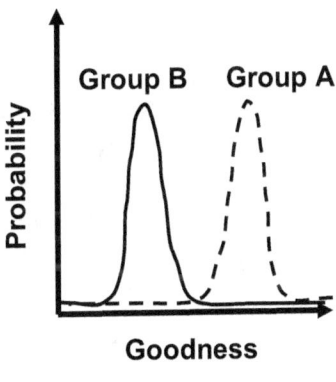

Figure 4.2

This modification of the definition of goodness is also used by the Christians to declare that all heathens will go to hell after the apocalypse, by the Muslims to declare that all infidels will share that fate after *qayamat*, by Catholics to declare that Protestants are destined for hell, the Evangelical churches to proclaim the doom awaiting Catholics, and so on. This formula is used by various other groups to declare that none except the members of that group shall achieve rapture, salvation or bliss.

The goodness redefinition approach used to be very effective in the medieval times. Unfortunately (or rather very fortunately), with religious tolerance becoming more of a norm in current world, this nice clean solution can no longer be used equally effectively by a cult leader to motivate and persuade his followers.

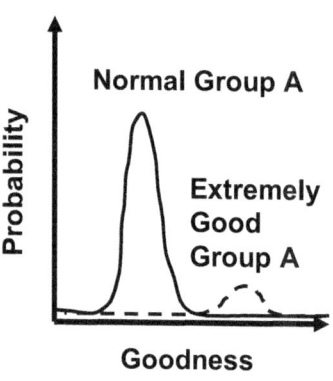

Figure 4.3

So, the smart leader of modern time uses the statistical distribution of goodness in a more sophisticated manner to create the illusion of goodness being distributed as shown in Figure 4.2. Suppose he looks at the distribution of goodness in his own group, i.e. Group A. There will be a small set of people in Group A which will be at the very top end of the scale of goodness. These are the people who are always kind and considerate, help out others, and are good model citizens in most respects. If you divided the followers of Group A into the set of people who are extremely good, and the followers of Group A who are normal, you will get a

PDF distribution of these two subsets of the groups as shown in Figure 4.3. Since the extremely good people in Group A will be a fairly small proportion of the overall population, there would not be an appreciable shift in the PDF of the goodness of the rest of the population.

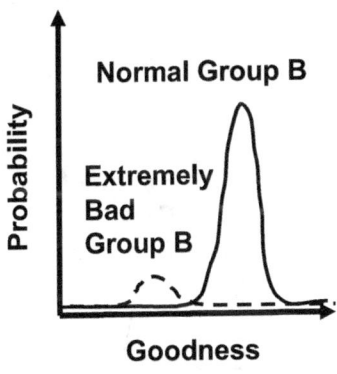

Figure 4.4

Now, the leader applies the same approach to the distribution of Group B. However, instead of looking at the extremely good subset of the population, the leader would look for that subset of Group B which is extremely bad. These are the people who are precisely what one does not want to be, people who commit heinous crimes like murder, child molestation, stealing and various other evil. Every large group has its share of such people, the folks who a leader would much rather not have in his or her following. If you do this subsetting on Group B, you will get a PDF of goodness for Group

B as shown in Figure 4.4.

Now all that the leader has to do is to compare the extremely bad subset of Group B with the extremely good subset of his own followers in Group A. One just ignores the normal group. After all, they are basically indistinguishable from each other. And when you do this comparison you come up with a very desirable shape of distribution as shown in Figure 4.5.

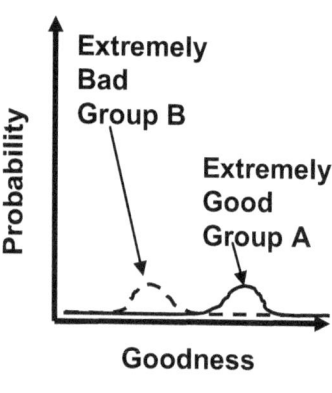

Figure 4.5

Now just drop the Extremely Bad and Extremely Good prefixed on the labels from the diagram on Figure 4.4, normalize the distributions, and you come up with the nice looking PDF distribution of Figure 4.2 which is what the leader of Group A wanted in the first place.

It is for this reason that you find a great deal of

attention paid to the evils of Catholic Church by the evangelicals. The Catholic leadership returns the favor in sermons and broadcasts to members within its fold.

It is not just the religious sects which use this nice handy tool of statistical manipulation with each other. If you look at the Republican and Democratic parties going at each other, you can find the same selective comparison process being followed in full script. Most of the traits you would associate with bring a good American citizen are independent of an individual's preference for a specific party.

As an example, let us consider the case of moral behavior among American politicians. For the purpose of this example, let us define moral behavior as adoption of practices and a life-style which will prevent the politician from earning the dubious honor of a listing on the ranks of sex scandals maintained by websites such as ranker.com. At the time of writing of this book, there were twenty-seven Republican sex scandals listed on the website, one less than the number of Democratic sex scandals. This comparable number of scandals would be natural to expect given the fact that moral behavior would have a comparable PDF among politicians of both the parties.

If you listen to the political spins of politicians belonging to either of the two parties, you would end up with the impression that the other party is full of immoral scumbags while the speaker's party is full of honest hard-working folks who have a high degree of moral behavior. While most politicians are smart enough to stop short of stating this in explicit terms, there are enough hints that convey this message.

You can repeat the entire discussion covering figures 4.1 through 4.5 with the horizontal axis being labeled as moral behavior and the two groups being Republican and Democrats and you can make out the script of the debates and discussions ranging among the political parties. The debate structure is very clear; each party will take the best among its group and compare it to the worst among the other group to show its superiority.

You can find the same script being followed to earn the battle of the minds in the seemingly endless conflict between Israel and the Palestinians. Each side accuses the other of inhumanity and cruelty. If you read the Arabic press or follow the opinion section on English channels of news source such as Al-Jazeera which are sympathetic to the Arab cause, you can go through the innumerable cases of despicable acts committed by the Israeli defense forces, destroying the lives and livelihood of innocent and defenseless Palestinian civilians in wanton acts of violence and unprecedented cruelty.

On the other hand, if you read through the Hebrew press, or English news sources like New York Times which tend to adopt a pro-Israeli stance, you will find elaborate coverage of the terrorists championing the Palestinian cause who send killer missiles into civilian homes, dress up babies as suicide bombers, and cause random acts of violence.

The basic script being replayed here is the same - compare the best of one group against the worst from the other group.

5. College Admission Tours

College admission tours are a rite of passage that most high school juniors undertake every year in the summer before they begin the senior year. Like the medieval pilgrims heading towards holy cities, they choose the colleges and universities of their liking, and pay a visit to those seats of learning. Some may shift the time for pilgrimage to spring break or other times, but a large fraction of American high school population takes this journey at one time or other. Parents usually accompany their children on these tours, but on some occasions a close relative or family friend may also be called upon to help in this quest.

Due to a scheduled work trip to the Boston area for work, I ended up attending the college admission tour for MIT with a young relative. The Massachusetts Institute of Technology is one of the best technical colleges in the world, and only admits the best and the brightest in the United States, combined with a very select program for international students. Any college going kid who is interested in a science related discipline would consider it a great honor and privilege to get admission there. I was sitting in one of their information sessions, mentally prepared to be dazzled and wowed by the tremendous technical brainpower that is packed in the institute's faculty and students.

There was some mention of the great technical achievements of the college, but it was almost in passing. I was surprised to discover the college admission officer and the young tour guide who took us around the campus dwell for the majority of the information session on the fun side of MIT There were elaborate discussions on the tradition of hacking[4], funny anecdotes of bridges getting measured in lengths of students, contraptions designed for instant delivery of food to the dorm-rooms, and other frivolities that the students indulged in. After the tour and information session, I was almost convinced that MIT is a big party school where students do all kind of crazy funky things.

On the way back, I was just wondering about this surprising side of MIT that I had just discovered. Based on my interactions with their graduates in the field of engineering, I was under the solid impression that MIT graduates great scientists and engineers, for the large part really serious students with little interest in frivolities of life. On a basis of the interactions I have had with MIT graduates, I would consider less then 1%

---

[4] A hack in MIT. jargon is a harmless but ingenious prank undertaken by students.

of them as being those likely to be indulging in fun and frolicking. Almost all of the MIT graduates I have met are geniuses, and they all seem to fit well Newton's adage that genius is 99% perspiration. They are all hard-working and deeply technical people. The concept of MIT as a party school just did not match with my personal experience.

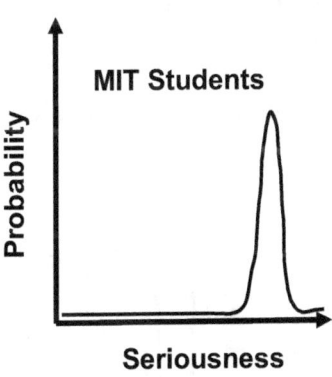

Figure 5.1

Let me explain the difficulty I was having in terms of statistical distributions. Suppose we define a metric called seriousness, which is low for students who like to party, fun and frolic, and high for students who are very studious and industrious. If you take a given class for any year at MIT, the PDF of seriousness among the MIT students would look something like the graph in Figure 5.1. The PDF would have a peak occurring towards the right-hand side of the graph, showing that most MIT students have a very serious attitude towards studies, with a few being fun loving and frivolous.

The impression that was created from the admission information session was that the distribution was of the nature shown in Figure 5.2. In this case, the peak of the distribution is markedly towards the left hand side, showing that the MIT student population consists primarily of students who love to have fun, play pranks and interesting hacks on the system. The information session never put it in these specific words, but that was the impression the audience was left with.

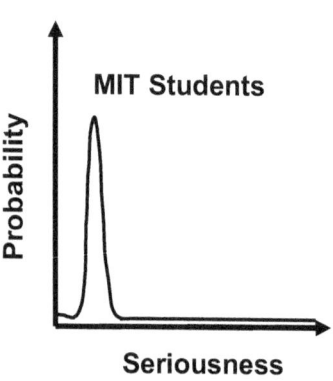

Figure 5.2

The genius of the MIT admission coordinator was to take a population with the characteristics of Figure 5.1, and make it look like Figure 5.2.

On the way back after the admission tour, I reasoned that there probably was some marketing logic behind the portrayal of MIT in this manner by their admissions coordinator. Everyone knows about the great technical merit of MIT and how competitive admissions are, so the admissions coordinator probably opted to highlight on the fun aspects and lighter side of student life to put forth a more welcoming image of the college.

But that rationalization still left me struggling with one issue. How did the admission officer manage to give such a different impression, represent a population that looked like Figure 5.1 into a population that looked like Figure 5.2 without stretching the truth in any manner? All the anecdotes, instances and hacking stories narrated in the information session were all very factual and verifiable.

And then it dawned upon me how the MIT admissions officer had done it. Once again, it was an adroit manipulation of information using the nature and characteristics of statistical distributions. It was a stroke of genius, and my respect for MIT ingenuity was doubled immediately. The MIT trip showed me how one can create an impression of a distribution which may be the exact opposite of what the reality could be.

Let us take the MIT class of 2011. It will have a distribution of students which will be similar to that shown in Figure 5.1 when you take into account the seriousness of the students. So, if you take into account the size of student body which is graduating in 2011, you may probably have a couple of pranks or hacks done by the students in this graduating class. As a matter of fact, the graduating class of every year would have a couple of interesting pranks or hacks. The specific incidents of one specific year may or may not be noteworthy, but if you look over the history of MIT which has been around for about 150 years, you get a sizeable number of such incidents throughout its history, somewhere in the order of 300-400 incidents assuming two or three incidents every year. Now, all that the admissions coordinator has to do is to pick some ten of these incidents to include in his/her portfolio to present to prospective candidates.

Each year, there are probably tens (or even hundreds) of noteworthy technical achievements of MIT students. The admission coordinator now has a choice of a few thousands of such achievements to pick from. Picking out four or five such achievements from the vast history of a great college is fairly easy.

Now, the admission officer simply has to present some ten hacks with a couple of minutes on each, and four or five technical achievements with a minute on each. The net impression on the audience is that of a student body which spends a lot of time on fun filled pranks, and incidentally also happens to do great technical things on the side – reversing the role of the two in reality, where students spend a lot of time doing great things, incidentally managing to do a couple of pranks on the side.

And depending on how one decides to emphasize the different activities, one can make a serious school look like a party school, or make a party school look like a very serious school.

A statistician will probably call such manipulation a biased form of sampling. We are taking a distribution, and then selecting elements from that distribution that are chosen in a way to cause a distorted view of reality. However, such manipulations are fairly easy to do when a distribution consists of a very large number of samples.

College admission tours are not the only ones who can try to subsample data selectively to try to shift the situation to look like the exact reverse of reality. One can even overlook the manipulation done by the admission guides; it is after all their job to market their colleges as effectively as they can. However, selective sampling can be used to make convincing arguments in various other cases. One such situation deals with the case of nuclear non-proliferation.

Ever since the world discovered the destructive potential of nuclear weapons, the United States has been at the forefront of trying to prevent proliferation of nuclear weapons. While there are many who have questioned the motives of the United States in trying to promote such proliferation as being selfish, few would doubt that the United States takes every possible precaution to prevent the spread of nuclear weapons and specially having them fall within the hands of stray terrorists who are very likely to use it against the United States.

On May 3rd 2010, President Ahmadinejad of Iran made a passionate speech in the United Nationas. Part of the speech described how the United States is one of the worst nuclear proliferators with a hypocritical and irresponsible behavior pattern. He put the following facts on the table – The government of the United States that is the biggest culprit in the production, stockpiling and use of nuclear weapons The United States has never respected any of its commitments. In the past decades, the United States had most of its wars with its friends. Some member states of NPT were threatened to be the target of a pre-emptive nuclear strike. The United States has always diverted public attention from its illegitimate actions bringing into focus misleading issues. Meanwhile major terrorist networks are supported by its intelligence agencies. Those who exploded atomic bombs killed hundreds of thousands of people and leveled to the ground two cities, are regarded as the most hated human beings in the history. His summary thesis, delivered at a Nuclear Summit in Teheran in April of the same year, was that the United States is "the sole nuclear offender, which falsely claims to be advocating the nonproliferation of nuclear arms while doing nothing substantive for this cause."

Since the statement was made by an individual known to be virulently anti-American, it was dismissed and ignored by most of the mainstream media as yet another instance of Anti-American rant from the president of Iran. Nevertheless, one has to marvel at the sophistication with which the argument was constructed, and the selective selection of events to create an impression that is totally opposite of reality.

Another instance of the same principle is used by people who selectively quote verses from Bible in order to support their position. The Bible is a lengthy piece of literature, with about 260 chapters and close to 8,000 verses when all of its various books are taken together. With so many verses, you can selectively sample verses to support almost any topic under the Sun. You can pick a cause, no matter how despicable, and manage to find five or ten verses in the Bible that can be strung together to demonstrate that the Bible justifies that cause. If you do not believe it, keep in mind that there were many people who quoted biblical passages to support slavery a couple of centuries ago.

As another example of selective sampling, let us consider women's rights under Islamic Sharia law. In this case, I will refer you to a nice treatise on this topic by Dr. Zakir Naik. Dr Naik is an Islamic scholar who takes a position that "Islam promotes several types of women's rights," while "Western society has degraded them to the status of concubines, mistresses, and society butterflies who are mere tools in the hands of pleasure seekers and sex marketers."

In defense of his thesis, Dr. Naik has listed out a series of various rights that are listed for women in Koran and Islamic Sharia law. These include social, political and economic rights. Dr. Naik also points out the fact that these rights were conferred in Islamic laws 1,300 years before Western society gave equal rights like voting to the women[5].

Dr. Naik is absolutely right in all of his facts, and totally wrong in his conclusions. He lists nine rights that are allowed to the women in Islam, and completely ignores many more rights that are available to women in modern western society, and denied to women in Islam. By this selective sampling, he is able to make a seemingly cogent and logical case for things that are blatantly false.

And if you think nobody in their right mind would believe the specious arguments of a person like Dr. Naik, keep in mind that he has several enthusiastic followers, and was considered enough of a threat by the UK government to deny him a visa to enter that country.

---

[5] See URL http://www.witness-pioneer.org/vil/Books/ZakirNaik/ZakirNaik_GenderIssue.htm

6. Walking with Dinosaurs

Although it may appear somewhat anachronistic, there is still an ongoing debate between the proponents of the theory of evolution and the theory of creationism. There are strong opinions on both sides.

Evolution is the theory which states that animal and plant life evolved by means of genetic mutations and survival of the life-forms that were best adapted for the environment. There is good evidence for evolution on the basis of genetic makeup of different species, and various fossils that have been found so far. Although there is room for improvement and some unanswered questions in the theory, it is accepted as the best known explanation for the origin of life and different species by the majority of scientific community. The room for improvement is not much different than the search for a grand unified theory in the field of physics, where scientists are still looking to improve the current state of knowledge.

Creationism is the theory that posits that all intelligent life was created by a super-natural entity. The proponents of this theory are generally inspired by a strong belief in Christianity. Although creationism is not accepted as a sound scientific theory by the majority of scientific community, it has a strong group supporting its cause politically and financially.

The debate between evolution and creation is an acrimonious one. We are not going to be sorting out the debate between evolution and creation in this book, since that is well beyond the scope of what can be sorted out by means of statistical distributions. However, among the different topics that creationists and evolutionists disagree about, there is one topic which can be understood using statistical distribution.

That topic deals with the issue of co-existence of dinosaurs and humans. Based on the fossil evidence that is available, most evolutionary biologists believe that human beings lived in an era ranging from approximately 65 millions years ago to 245 million years ago. Similarly, hominids (or ancestors of humans) are believed to have been present on Earth at most million years ago and modern humans appearing on the scene less than 500,000 years ago. As far as the current thinking in evolution goes, humans and dinosaurs never co-existed together on Earth.

On the other hand, creationists (or at least a subset of creationists) believe that dinosaurs co-existed with human beings. The rationale of this belief is tied to the biblical reference of God creating the world and animals in a period of six days. If that was indeed correct, then dinosaurs and humans should have co-existed. The creationist museum in Kentucky, operated by a group of creationists to promote their world-view, has exhibits showing the co-existence of humans and dinosaurs.

It is unclear how the co-existence of humans and dinosaurs settles the debate between creation and evolution. Life could have evolved in a manner that the dinosaurs and humans co-existed, and a supernatural creator could have made humans after all the dinosaurs had died. Despite its seeming irrelevance, this topic has become a bone of contention between the evolutionists and creationists. One possible reason from a creationist perspective would be that if dinosaurs did co-exist with humans, the inferences made from fossil records (which support the concept of evolution) could be called into question.

It would be interesting to see if we can use the concept of statistical distributions to shed some light on this aspect of the debate. On this specific question, statistical distributions would seem to support that the creationists may be right, and that dinosaurs and humans could have co-existed.

The oldest known humanoid fossil at the time of writing this book is the skull of a ape-like creature discovered in Chad in 2001 estimated to be somewhere between 6 and 7 million years old. The oldest known fossil of modern man (Homo Sapiens) is the Heidelberg man estimated to be about 600,000 years old. The total number of human or humanoid fossils that have been found would probably number in a few thousand specimens.

Similarly, the youngest known dinosaur fossil is about 65 million years old while the oldest known dinosaur fossil is about 245 million years old. The total number of dinosaur fossils found in this period number in a few tens of thousands.

Because of the age of the discovered fossils found thus far, the evolutionists take the conservative opinion that dinosaurs lived in a period starting about 245 million years ago and ending about 65 million years ago. That time period does not overlap with the time when humanoids existed, at most 7 million years ago.

However, it is worth noting that the limits on the range are not hard bounds on the life-span of either of the two species. We definitely know that dinosaurs existed between 65 and 245 million years ago because corresponding fossils have been found. However, tomorrow a fossil showing a dinosaur living 25 million years ago may be found, and the next day, one may found a fossil which shows a dinosaur that lived 400 million years ago. In that case, the age of dinosaurs will be modified to the range of 25-400 million years ago. The time-period when dinosaurs lived and when human beings lived are not upper bounds, they are the minimum bounds established by the current fossil evidence.

It is conceivable that one may find a humanoid fossil which is 70 million years old. In that case, humanoids and dinosaurs could have potentially co-existed. In other words the fossil evidence does not discredit the creationist claim that humans could have existed with dinosaurs.

And it gets worse when we introduce the concept of statistical distributions. You must have noticed that the number of fossils found for humans and humanoids are a very small fraction of the total number of humanoids that have lived. Currently, we have about 7 billion humans on the planet. If you count all the humans and humanoids that must have lived over the past 7 million years, the number of humanoids would easily be a few tens of trillions. However, we have only discovered a few thousand fossils, which is a negligible fraction of the total humanoids that ever lived. This is not unexpected. Very few fossils will survive for a long time, since finding a fossil requires that an animal die in or be moved after death to a location where the remains would encounter conditions suitable for fossilization. One would expect a similar situation to hold true for dinosaur fossils, that the number of discovered fossils is but a negligible fraction of the overall number of dinosaurs that lived.

Suppose we plot the probability distribution for the existence of a dinosaur over time. On the horizontal axis, let us plot the time starting from about 300 million years before the present time. On the vertical axis is the number of dinosaurs that were living at that given time. However, we scale the vertical axis so that the area covered under the curve sums up to one. This gives the probability that a randomly selected dinosaur that ever lived on Earth was living during the time shown on the horizontal axis. If this appears to be too confusing, just consider the curve as a scaled down version of the number of dinosaurs living at any time. This curve would have to begin sometime before 245 million years in the past and has to end past the 65 million years in the past. At that time, we can assume that at least one dinosaur was alive on the basis of fossils that were found.

In essence, what we are trying to do is to determine the right distribution when there are only some samples available from that distribution. The problem can be illustrated by the diagram 6.1. It shows a few points, marked by stars, which are the dates of known fossils. When multiple stars exist on the same vertical axis, multiple fossils are found at about the same time. If there are multiple stars at any given time, we know the probability value of the PDF would be higher at that point, and where there are less stars, the corresponding PDF curve will be lower. We need to find the right distribution that provides a good fit for the samples that are collected.

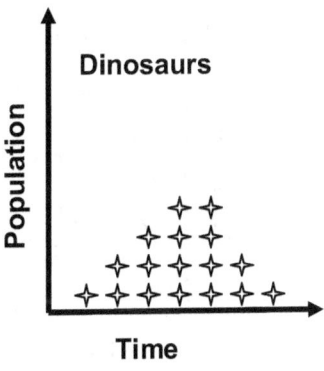

Figure 6.1

The challenge here is that the same set of points can fit several distributions quite effectively. Figure 6.2 shows three such distributions of dinosaur population that would be perfectly legitimate explanation of how dinosaur population fluctuated during the different years, and is consistent with the observed fossil record. They all show a higher population when a large number of fossils are seen. Line A assumes that the dinosaurs' population started when the first fossil was found, and ended when the last fossil was found. That is the most conservative estimate one can make in regard to the population of dinosaurs. The other curves (B and C) assume that dinosaurs lived past the age where fossils were found.[6]

---

[6] There are some other curves that can also be drawn, but we are

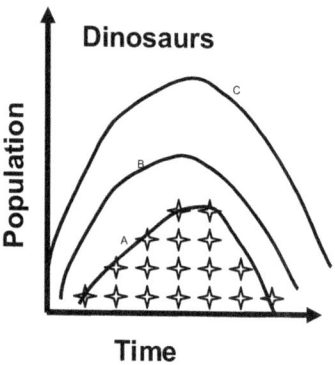

Figure 6.2

Let us consider the fossils of dinosaurs, where we know that the oldest fossil is 245 million years ago and the most recent fossil is 65 million years ago. The tricky point is determining how far the distribution extended before 245 million years and after 65 millions years in the past. The answer depends upon the expected number of living animals that are likely to generate one single fossil. If we assume that every tenth dinosaur became a fossil, then the curve would extend only a little bit past both definite boundaries. On the other hand, if only one in ten million dinosaurs became a fossil; the PDF would extend for a significant time beyond the two limits.

The expected number of dinosaurs that would have

---

making the assumption that the population changes are gradual.

become a fossil is hard to determine, so let us try to make a reasonable guess. But we know that as we move away from the boundaries known, the probability of the dinosaurs existing will become much less. Therefore, the distribution of the population of a dinosaur would look like that shown in Figure 6.3.

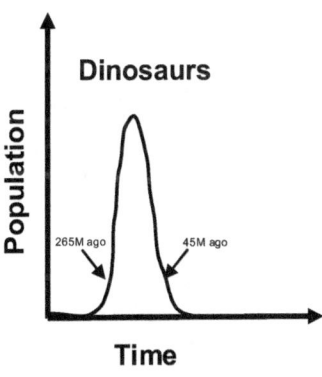

Figure 6.3

We can repeat the same exercise for the fossil records found for humans. The oldest fossil of the modern human is about half a million years old, and humanoid fossils were found almost seven million years ago. We know that the population of modern humans is growing, so we are in the rising side of the population distribution one would expect to see. One can use logic similar to that in the case of dinosaurs to estimate what the distribution would look like for human population. Since we don't know the actual ratios of dinosaurs to people, we will again use the scaling principle to make the area under the human population curve come out to be about a unit.

On overlaying the two, we will find a situation like that in Figure 6.4.

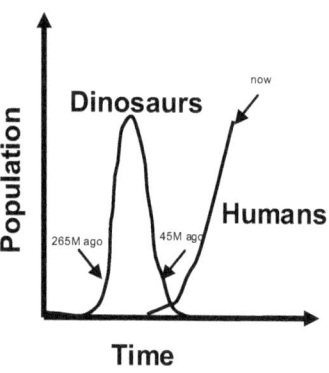

Figure 6.4

From Figure 6.4, you can see that there is a small overlap between the times when dinosaurs can be expected to exist and when human beings can be expected to exist. So, it is possible that dinosaurs and humans co-existed.

The real question is about the probability of that existence, i.e. how big is the overlapping area when compared. It would depend on the assumptions you make, but as you can determine from Figure 6.4, it is not likely to be very high.

So, the creationists are right in this debate, or more precisely, they might be right that dinosaurs and humans co-existed on Earth. However, the burden of proving that the two species actually did co-exist lies on the shoulders of the creationists. One way they can do it is by finding the fossil of a humanoid that is older than the fossil of a dinosaur. It should not be the job of evolutionary biologists to disprove that the two species did not co-exist.

7. Unequal Opportunity

Although the United States is the biggest bastion of capitalism and free enterprise, most people understand that unbridled capitalism can have some undesirable effects. One of the problems of unbridled capitalism is that it puts the under-privileged of society into a vicious cycle. In order to get to a better social status, they need to have access to resources such as education and training but they can not afford to do it because of financial reasons or prejudices existing in the society. So, the under-privileged need some support to break out of the vicious cycle. It is with this goal in mind that affirmative action and equal opportunity programs have been instituted. A large fraction of the American population would agree that the objectives and end-goal of these programs are worth achieving.

When it comes to implementation of these objectives, care must be exercised. An improper implementation of an affirmative action program results in exactly the opposite result. A classic example of such an inappropriate implementation is apparent in the large company that employs one of my close friends. The friend lives next door, and we meet almost on a daily basis. Through various interactions during a long sequence of backyard barbecues and shared dinners, the story in this chapter has been pieced together.

This chapter tells the story of a company in which affirmative action was introduced. Unfortunately, the net result was not good since the executives of the company forgot some basics about distributions in the society which were outside their control.

My friend works in the engineering department of a large company. For various reasons, engineering does not have a large representation in the workforce from minority groups such as blacks or Hispanics. There are also very few women in engineering. As a result, there is an increasing concern among the executives of the company that the engineering department may be biased or prejudiced against minorities and women. If such a bias existed, that would be very understandably a cause for serious concern. My friend had gotten an engineering degree from one of the liberal west-coast colleges, and always tried to promote women and minorities in engineering through a variety of programs.

Factors such as race or gender are irrelevant to one's capability or merit as an engineer, or for most of the departments which many modern companies consist of. In an ideal unprejudiced world the distribution of irrelevant factors such as race or gender in any department with a large number of employees should be roughly the same as the distribution of the same factor in the general population. With this thought in mind, the executives of the company ran an audit of all departments for hiring and promotions of under-represented minorities and women. While most of the departments passed the audit, the engineering department failed in a spectacular manner.

Let us represent the situation in terms of statistical distributions. Since race is a discrete quantity, the use of a discrete PDF represented as a bar graph would be most appropriate. We will also divide the population into only two groups instead of multiple races. The first group (Group A) consists of under-represented minorities and women. The second group (Group B) consisted of white males and other well-represented males such as Asians. In the general population, the distribution of the groups is somewhere along the lines of about 55% of population belonging to group A, and 45% of the population belonging to group B[7]. The distribution of the groups in the general population is shown in Figure 7.1.

---

[7] Women are part of group A, and make up 50% of population. That is why group A is larger, even though it consists of under-represented minorities.

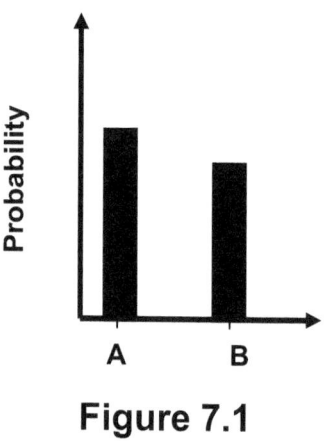

Figure 7.1

On the other hand, when the audit was taken of the engineering department, the distribution of the groups appeared as shown in Figure 7.2. About 80% of the engineers belonged to group B.

The executives of the company declared this situation to be unacceptable. As my friend explained to me, Human Resources of his company took the position that the statistics indicated existence of potential bias and prejudices in the engineering department's hiring practices. HR also was concerned that the mismatch of demographics may result an investigation of the company's hiring practices from the United States Government, and the lack of diversity in engineering department may result in significant punitive damages. As a result, all of the engineering managers were given directions to try to improve the hiring of members belonging to group A.

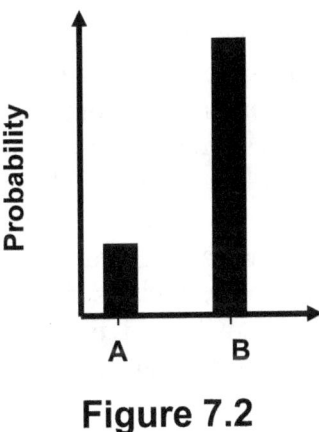

Figure 7.2

After a couple of years of hiring under this new practice, the company again conducted an internal audit. The mix among group A and B in the engineering department had hardly budged. Human Resources and company executives were now genuinely worried. They suspected that prejudices among the engineers who belonged to group B might be preventing the hiring of group A candidates. More drastic measures were needed to be put into place. Explicit targets were set for hiring new employees belonging to group A.

Under these hiring practices, my friend got a project for which he needed to hire six new employees. He was quite excited about the project, and its details were the subject of many shared meals. He was given approval to hire them subject to a soft guideline that he try his best attempt to fill half or more of the new slots by candidates belonging to group A. When recruiting the new employees, he tried hard to find suitable employees of group A. Despite all the attempts, he could only find one person belonging to group A who did well in the selection process despite interviewing several such candidates. On the other hand, several candidates in group B did very well in the interview process. Over a shared dinner one day, he told me that he had decided on his final slate of candidates for the project, with one member of group A, and five members belonging to group B. He was scheduled to meet his boss the next day.

The next day, we were barbecuing together, and my friend did not look very happy. On a little bit of gentle probing, he flew into an emotional outburst with some very rather harsh words against his boss. His boss had not approved the recommended list of candidates. My friend had been given a rather long and painful lecture on how to identify and eliminate prejudices in hiring. He also was told in no uncertain terms that he had basically two choices: he could either come back with a list consisting of half the candidates belonging to group A, or not have any people to hire at all on his project.

I asked him what he would do, and he shrugged his shoulders. His project deadlines were slipping, and there was a dire need to hire new engineers. He said he would choose the less of two evils, and just hire two more members of group A, even if all they could do was breathe and warm their chairs. At least he had four capable candidates, and his project could benefit even if a couple of people were not that competent.

At a shared dinner a few months later, he was looking much less sanguine. Over the initial glass of wine, he revealed his frustration. He had to give the annual performance assessment of engineers in his team, and two members of group A were ranked as the bottom performers in his group. His boss had given him a stern lecture on how it was unacceptable to have members of group A at the bottom of the list, and could be viewed as a sign of bias. He also told me that he had resolved not to switch the performance ranks around since that would be unethical.

What happened about two months later was even more shocking. Our families were supposed to have dinner together, and my friend said that he may get a few calls as he tries to find suitable engineers for a new project that he was leading. I asked him if these were new hires, but it turned out that he was just looking to find people working on the project from other engineers already available at the company. Some of those engineers would be calling him back even though it was past the work day. He ended up fielding three calls during that dinner as he was doing his selections.

Two of the calls were relatively short, while he spent a lot of time grilling the engineer on the third call.

I remarked on that and he just shrugged his shoulders and said, "Oh! That is because it's an engineer belonging to group A that I don't already know."

It sounded like a somewhat racist statement for someone with his background and liberal views. I couldn't resist pulling his leg by commenting on that fact. At this, my friend suddenly grew very serious, pulled a table napkin over to him and drew the diagram of Figure 7.3.

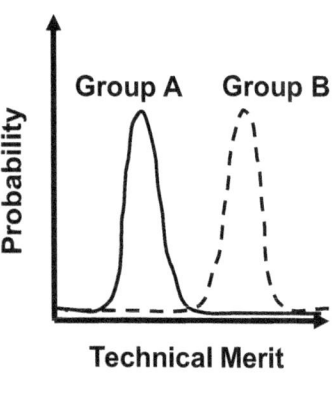

Figure 7.3

He reminded me that he had to hire two engineers belonging to group A even though they had less technical merit than other candidates. Other managers at his level were also making similar choices, and the percentage of members of group A was gradually increasing in the engineering department. However, since members from group A were being hired with a less stringent criteria, the distribution of technical capabilities among engineers of group A and group B had become like Figure 7.3.

As projects came into the engineering department of my friend's company, they were assigned to a lead engineer, who helped in recruiting the right people from across the engineering team. Most lead engineers were also managers, and they staffed the projects from existing engineers from different groups managed by them or their peers. Because of the questions about technical merit about some of engineers belonging to group A, each project leader exercised extra caution when asked to staff a project with any engineer of group A.

My friend explained to me that he was just being cautious. He pointed out the situation with the three members of group A he had hired in the past, one of whom he had selected initially based on technical ability and two who did not meet his criteria for ability but were hired nonetheless. Since there were several such instances of hiring in the company, he had to exercise extra caution when taking anyone from group A he did not personally know onto his project. He had no idea if they were hired based on technical competency like the person in his initial list, or like the other two he was forced to take into the team in order to meet the mandated quota.

I pointed out to my friend that this was a definite case of biased behavior against people of group A. It was an implicit assumption that any person of group A was not technically competent. I also ran the math for him. His company had 20% group A engineers before the new hiring policies came into effect. In the years after the new practices came into effect, the total engineer population had increased by 20%, with 30% of the new population belonging to group A. Even if only 33% of new employees of group A were competent, the majority of engineers in group A would be technical competent. The distribution among the engineers would look like that in the table below.

Time	Group A Total	Group A (Capable)	Group A (Incapable)	Group B
Initial (of 100)	20	20	0	80
Final (of 120)	36	24 (20+4)	8	84

Of the total employee population of group A, two-thirds (24 out of 36) were capable engineers while only one-third would be ones with questionable capabilities. Yet everyone in group A was being viewed as being technically incompetent just because of a small minority. I pointed out that this was a blatant case of bias, and definitely not in line with the liberal views I had come to expect of my friend.

My friend just smiled in a weary manner, and pointed out that while he had to do an extra investigation to determine that an unknown engineer belonging to group A did not belong to the one third of that team, he did not need to do much investigation for members of group B. Most of those members would be technically competent. If it be biased behavior, so be it. The success of his project was more important than adopting an unbiased attitude.

The hiring policies instituted by my friend's company had made the percentage mix among the two groups look better on paper. However, in practice, the company had resulted in an organization where the managers and project leaders had adopted a highly biased view of the existing population. Instead of making the group of an engineer an irrelevant factor, the group of an engineer had become a very significant factor to consider in routine operations like staffing. The transformation of my friend, who used to be fairly unbiased person, to a biased person investigating engineers on the basis of their group, was the exactly opposite effect that any rational equal opportunity or affirmative action program ought to aim for.

One particularly interesting fact here is that no single individual can be blamed for this transformation of a relatively unbiased group into a strongly biased group. The company executives and Human Resources had the best of intentions at heart. My friends' manager was simply trying to enforce the guidelines that came from the top. He and other managers and project leaders in the company were trying to adopt the most pragmatic approach to get their work done. But the net result was the development of a tendency where all members of group A were implicitly assumed to be worse, unless one happened to know a specific individual. Even the existing members of group A acted with this biased approach.

The net situation was very similar to another racist idea that many of us used before the advent of GPS devices. We had a simple adage for staying safe if you were driving at night in a big city and needed to get directions. We would pull into a gas station and if the clientele had some whites, we will get down and ask for directions. But if the clientele was all black, we would discretely drive away. At one time, I was driving with a black colleague in Atlanta, and we lost our way. My colleague used the same racist criteria to decide which gas station to stop and ask for directions. Regardless of our own race, we were both using a racist and prejudiced approach. And we did not care that we were being racist because being safe was more important to us than being unprejudiced.

Why did all the well-meaning policies instituted in my friend's company go wrong? Why did normal people turn into biased people? One possible explanation lies in the nature of statistical distribution of technical skills among different groups.

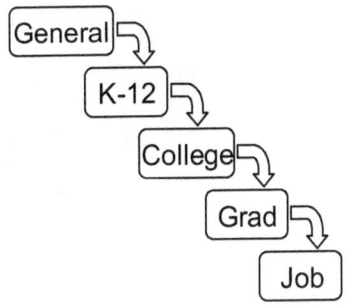

Figure 7.4

Engineering is a highly specialized profession. A typical engineer is drawn from a group that goes through several stages before becoming an engineer. Those stages are shown in Figure 7.4.

The first block is that of the general population. Some members of the general population graduate high school. Then they go to college. Some of them go on to graduate studies, and finally the engineers from the graduate school end up joining the job pool.

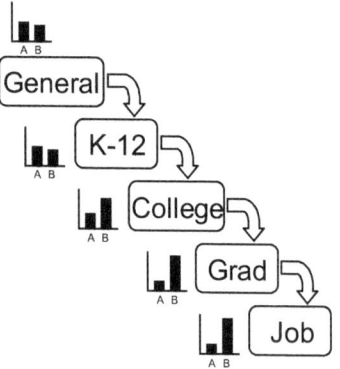

Figure 7.5

In each of these stages, there is a different demographic mix between group A and group B. If we want the demographics to look the same between the first stage (General Population) and the final stage (Job Pool), we can not ignore the demographic changes as they happen among the different groups.

Statistics about the demographics of different places are available readily on the web from the United States Census data, and various other agencies. As a representative sample, this book has used the statistics from the state of Indiana, using its general demographics, its high school demographics (available from Indiana DoE website), the undergraduate demographics of Indiana University (available from its admissions office) and the Ph.D. program in engineering disciplines from the same information.

The demographic mix among the different stages is as shown in Figure 7.5. Each stage is accompanied by a discrete distribution of members among group A and B.

As a part of the general population, the state of Indiana is about 85% white, 2% Asians, and 13% blacks, Hispanics and American Indians. Among this population, roughly half of the population is female. For the purpose of engineering organizations, we had defined group A as women, blacks, Hispanics, and other minorities and group B as male whites and male Asians. This gives the demographic as roughly 56% group A and 44% group B.

If we now look at the statistics of high school graduation in Indiana, the percentages are relatively the same, about 55% for group A and 45% of group B. However, when we look at the statistics for college undergraduate enrollment in engineering, the numbers shift significantly. Although the overall undergraduate population at Indiana University is about 51% group A, and 49% group B, when you look at the undergraduate engineering enrollment, the number changes to about 40% group A and 60% group B. At engineering related graduate disciplines at the same University, the mix is 15% group A and 85% group B.

Although the numbers used here are for one specific university, it seems reasonable to assume that a similar shift in demographic would be seen if one collected statistics across all of United States Universities. In that case, the original demographics between group A and group B in our company was about the same as one could expect. If about 15% of the job applicant pool belongs to group A, it would be reasonable to have about that percentage mix in any company hiring from that group. In any company not employing prejudiced hiring practices, one would expect to see 10% -20% of their employees as belonging to group A.

The mistake made by the executives of my friend's company was in trying to force a statistical distribution in the engineering division which would be the same in the general population. They ignored the shift in demographics that happened in different stages of the training that leads to an engineer. This caused them to ask for a distortion of the demographics of engineers which was just not viable.

The stresses and biased behavior that resulted in the company were an outcome of the unrealistic goal. If the executives had based the target distribution on the statistical distribution of graduating engineering population, they would have not set an unreasonable target, and introduced biased behavior among engineers.

The same principle probably applies to college admissions where a certain amount of diversity is desirable in the student body. However, when the emphasis on diversity takes on an unreasonable shift in the natural demographics of the student body, such policies may end up having the exact opposite effect of what they were designed to do.

8. Summarizing it All

Until now, we have looked at the various misconceptions, lies and erroneous statements which can be fixed by understanding the statistical distributions that underlie those statements. In this last chapter of the book, we will summarize the basic mistakes that people make, and which have been reviewed in a different manner throughout the book. Those mistakes are:

Confusing the properties of an individual with that of a group:

An individual has a property which is usually a unique value, e.g. the height of a person. All individuals belong to one or more groups, depending on the definition of a group. The same property will have a distribution when measured over different individuals of a group. An observation made about a group does not necessarily transfer to that individual, nor does the property of an individual necessarily transfer to that of the group.

Just because Bernie Madoff, an investor banker, was a crook, we can not conclude that all investment bankers are crooks. Nor does the fact that most priests are virtuous and moral mean that the priest in your local church is necessarily incapable of evil-doing.

You can't transfer relative positions of distributions among two groups to relative positions of specific individuals belonging to a group.

This is a continuation of the theme that the properties of an individual can not be transferred to become the properties of a group and vice-versa. While the distribution of a group is influenced by the specific properties of individuals, you can not mix the properties and comparison of a group to that of specific individuals within that group.

When considered as a collection, two groups may be compared on the distribution of a property within that group. However, that relation will not necessary be valid for specific individuals in that group. As a group, whites have a higher average income than blacks. However, that does not mean that every white person makes more money than every black person. You can be pretty confident that Oprah Winfrey makes more money than the white high school principal of the local middle school. Just because men are stronger than women on average does not mean that every man is stronger than every woman.

When comparing groups of people, be careful of how the comparison is being done.

When groups of people are compared, it is useful to

understand how the comparison is being done. Since a scientific valid sampling of two populations is rarely available for a meaningful comparison, people create a subset of any two populations to compare against each other. If the sampling process is not properly done, you will come up with an incorrect result.

We saw an example of this in which we compared the subset of gifted Asian people against a large population of Americans to draw the incorrect stereotype that Asians are better at math and science than non-Asians.

An extreme case of comparing two distributions is to take the best from one group and compare it to the worst in the other group.

This is the favorite tool of demagogues who try to show the values of the best among their group with the worst of the group they are against. Whether it be leaders of one religious sect trying to rail against another sect, or one set of jingoistic people trying to put down another group of people, this is a convenient tool to use.

If you select a subset of values from a distribution, you can portray a group in a very different light.

When one is dealing with a large distribution, you can manipulate the set of people you use to distort the reality of the situation. This is a favorite tool of the marketing department of many companies. By properly manipulating the data that is being presented, one can make a tough technical college look like a party school, and the reverse. And with selective quotes, you can get support for any cause under the sun from the Bible. And using the same technique, you can show that women enjoy wonderful rights under the stringent yoke of Islamic Sharia law.

When someone tries to change the natural distribution of a property in a group, one should take the constraints and dependencies imposed by other environmental factors into account.

In this book, we saw an example of how well-intentioned measures that did not take the dependencies into account resulted in the exact opposite effect of what was desired.

❧⸙❧

I hope this book was an interesting read, and you have now gotten a good sense of statistical distributions, and how they can be used to catch the different lies and misconceptions that others may put in front of you.

❧⸙❧